ABOUT POLLEN

A fascinating survey of the richest source yet revealed of vitamins, minerals, proteins, amino acids, hormones, enzymes and fats. Author demonstrates how pollen, essential for all plant life and garnered by bees, is a perfect food and medicine.

By the same author

ABOUT COMFREY
ABOUT GARLIC
MUSHROOMS: NATURE'S MAJOR PROTEIN FOOD
ABOUT SOYA BEANS
ORGANIC GARDENING AND FARMING

With Alan Moyle N.D.

ABOUT KELP

ABOUT POLLEN

Health Food and Healing Agent

by

G. J. BINDING M.B.E., F.R.H.S.

THORSONS PUBLISHING LIMITED
Wellingborough, Northamptonshire

First published 1971
Second Impression 1973
Third Impression 1974
Fourth Impression 1976
Fifth Impression 1977
Second Edition, revised and reset, 1980

ISBN 0 7225 0660 0

*Made and Printed in Great Britain by
Hunt Barnard Printing Ltd.,
Aylesbury, Bucks.*

CONTENTS

ACKNOWLEDGEMENTS

I would like to express my sincere thanks for all the help given by:
MR HELIER CLEMOES
AB CERNELLE, Vegeholm, Engelholm, Sweden.
ORTIS, Eisenborn, Belgium.
MELBROSIN, Vienna, Austria.
BEE RESEARCH ASSOCIATION (Eva Crane, M.Sc., Ph.D., Director), Hill House, Chalfont St. Peter, Gerrards Cross, Buckinghamshire, England.
MR ALAN RANDERSON, Managing Director of Pollen Products Limited, London N12.

INTRODUCTION

Throughout the ages honey has always been accepted as one of the finest foods available to man. Centuries ago, when primitive races lived close to nature, they were compelled to hunt for food: honey and fruit juices were the only means of sweetening. As progress was made and ancient civilizations appeared, honey was highly regarded and praised in the work of authors and in records. It was mentioned in the Bible, Koran, Talmud and was widely valued by the Romans, Greeks and Egyptians. In fact in every land inhabited by the bee the same beliefs existed regarding the miraculous powers of honey as both a food and medicine. The ancient Greeks named it, 'The Nectar of the Gods': writers and learned men of many civilizations believed this nectar to be a wonderful food and storehouse of medicinal goodness. So for centuries the power of this natural food was simply accepted. It is only in the last few decades that scientists have been able to prove conclusively that honey is not only a perfect food but also a giant germ killer in which bacteria simply cannot exist.

Although sugars have replaced honey as everyday sweeteners in most countries, bees still present man with their fine nectar. In an ever changing world honey is as good to-day as ever, being one of the few things that the atomic age has passed by: its goodness is there for one and all to sample. Man also made use of other fine products of the hive including bees-wax, honeycomb, and eventually, royal jelly. All these good things are produced unfailingly by the little bee year after year, as surely as winter follows summer. The composition of this amazing food no longer remains such a mystery and far more is known about the life of the bee. Man today can answer nearly all the questions about honey, but in spite of this, some 2 per cent of its contents still remains a mystery and defies identification.

A great deal of research has been conducted into the use of pollen as both a food supplement and a medicine, and the discoveries made are proving the true worth of this remarkable substance.

It is said there is nothing new in the world and pollen is no exception. From the time blossoms appeared we have had pollen, but what is news about it is the fact that it is only during the last two decades that scientists and doctors have been able to prove its true value. Pollen came into the news over 30 years ago when a Swedish beekeeper, by means of pollen traps, was able to get the bees to unwittingly harvest pollen for him. This beekeeper's original pollen traps have now been superseded by a series of machines based on an electro-mechanical principle. A great advance from the primitive methods adopted by natives long ago, when kinds of fishing nets were used to obtain supplies floating on lakes and waterways. Thanks to the efforts made in Sweden and elsewhere, scientists have, during the past twenty years been able to obtain sufficient pollen to enable them to make tests on it, as a food supplement and for medicinal purposes. The result is that we have today ample proof of the health-giving properties of these minute grains.

The history of pollen, leading up to its acceptance as a health food, makes an interesting study, which, in spite of some 30 years progress, is in some ways only in its infancy. Sweden has played a leading role in making discoveries about pollen and now harvests crops mechanically. This demonstrates how progress has been made from a beekeeper with a few hives in 1950 to a large company supplying many varieties of pollen preparations for health, medicine and beauty aids to many countries in the world.

Bees, as soon as they appeared on earth, were instinctively aware of the nutritive value of pollen, centuries before man really knew anything vital about it. Today scientists and beekeepers are gradually catching up. This book tries to answer some of the questions. Much more is being learnt and in Sweden and elsewhere scientific and medical investigations proceed, so, one day man may know all the answers about the amazing world of the bee, in which pollen plays a vital and most important role.

CHAPTER ONE

WHAT IS POLLEN?

To every student of botany there is a clear understanding about pollen, what it is and the reason for its appearance. In simple terms it consists of the male germ seeds of plants, flowers or blossoms on trees, and as such plays a vital role in ensuring that plant life in the world continues. Pollen is really myriads of the tiniest of spores which appear in the anthers of blooms and in the cones of conifers. Single grains are invisible to the naked eye. To illustrate the prolific nature of pollen, one strobile of pine, or spike of ragweed, may produce 6,000,000 grains of pollen. Multiply this sort of figure by the number of blooms on a tree and we can understand why lakes or ponds near such trees as pine often have their entire surface covered with this flour-like substance.

Pollination

As plants flower, pollen is transferred from the anther of a stamen to the stigma of a pistil, and on reaching the ovary it brings about fertilization of the ovules and the growth of seed. Some plants are pollinated naturally by the elements. However, flowers bringing beauty to our gardens and containing nectar are mainly pollinated by insects, such as beetles, flies, bugs, wasps, moths, butterflies, thrips and last but most important to man, the bee. This is only part of the story of these minute granules, for research during the past sixty years has revealed a most remarkable thing about pollen. Each minute grain measuring about .002″ has the power of an atom in that it is able to fruit a massive tree. Frequently pollen alone is unable to set in motion this miracle of life and in this respect it is aided by masses of insects, especially the bee, in completing the cycle of life which is a process so baffling to man.

Miracle of Pollen

Until fairly recent times the miracle of pollen has been left entirely to the plant world but deep research has proved it to be a food from natural organic sources, free from added preservatives, colour or chemicals. Pollen contains the richest source yet revealed of vitamins, minerals, proteins, amino acids, hormones, enzymes and fats. As with honey and royal jelly, pollen is surrounded by a certain amount of mystery in that it contains other substances which so far defy identification. Intensive research into the fascinating subject of pollen has been taking place in twenty different countries. At the university of Arizona it was discovered that antibiotics were present in pollen. Within hours of being put on a slide, with a mixture of boric acid and agar agar to accelerate germination, pollen grains will have grown tails. Such germination can only be revealed under the microscope. This does not occur when pollen is taken as a food. So one remarkable thing about pollen is this built in life force which allows plants to reproduce

Natural Sources

We are very fortunate in that there are opportunities in most countries for gathering this amazing food well away from areas of intensive farming and the artificial fertilizers and chemical spraying that goes with it. Of the many wonderful things in nature, pollen perhaps stands supreme. It has been established that during the course of one year the bees of a single hive will gather and store therein up to eighty pounds of pollen. Having heard an expression, 'the bee's knees', I often wondered what it really meant. Well the manner in which bees collect grains of pollen on their hairy legs and methodically mix them with nectar, then fly back to the hive carrying the pollen in a ball between their hind legs, may have some bearing on this saying.

Study of Pollen

By means of microscopic examination, pollen experts are able to estimate the quality of grains. Shape and colour indicates the species of plant from which it came. Although it is not always possible to observe droplets of pollen at the base of the petals, it can easily be detected by its white, yellow or reddish powder, often coming off on our hands whilst handling flowers. Certain bees in the hive gather nectar to convert into honey, whereas others forage for pollen. The day's work is divided up within the hive as in an army barracks: the entire organization would do credit to the Brigade of Guards. Bees perform a remarkable, almost miraculous feat in the collection of pollen, for it is not swallowed into the carrying sac like nectar. The pollen is worked into a mass and carried by their rear legs, which are expertly designed by nature to carry out this task.

Varieties of Pollens

There are as many different pollens as there are various species of plants, flowers or vegetation from where a source can be obtained. As with various honeys, pollens show differing quantities of vitamins, minerals and other goodness. Whereas the content of any honey depends on the kind of blossoms from where the bees have gathered their nectar, so will pollens vary according to the mineral and other compositions of the soils in which the plants have grown. This is shown when, from a breakdown of pollens obtained from dandelion, clovers, and a variety of fruits, we find a highly nutritive and rich source of copper with good amounts of pantothenic, ascorbic and nicotinic acids. Also present in these pollens are mineral constituents, calcium, phosphorus, magnesium and iron, together with vitamins D and E.

Chemical Composition of Pollens

The vast number of pollens gathered by bees supply practically all the amazing ingredients of royal jelly, food of the queen bee and nourishment for larvae. Thus pollens in the form of royal jelly allow the hive to survive. Pollen builds up the bee and allows him to develop to full growth and permits the continuation of the species. Bee larvae fed on bee-bread or pollen increases in weight 1,500 times in six days. Pollen is stored in the cells of the combs by bees and finally becomes 'bee-bread' when fermented. Pollen works near-miracles for the bee and all tests have shown that it will do the same for human beings and animals alike. Pollen really is the bee, for this buzzing little dynamo is made of pollen, nothing more, nothing less. It makes up his wings, organs, muscles, glands, hairs and so on. All wear and tear of the bee's body is replaced by pollen, fuel for his endless excursions is supplied by nectar and honey which provides the enormous energy for him to perform his miraculous tasks. He is in fact pollen-propelled by honey, that wonderfully energizing food. A bee will visit up to 1,500 blooms in order to gather one sac full of nectar. During a lifetime of up to eight weeks a bee will produce something like half a teaspoonful of honey. In order for one pound of honey to be stored in the hive, some 3,000,000 separate blooms have to be visited. Exceptional colonies have been known to store as much as two pounds of honey in one day.

Vitamin Content of Pollen

The nutrition contained in pollen is of vast importance to bee keepers, farmers, fruit growers and, in recent years, to animal breeders. It has also now been proved, especially in Sweden, that pollen is equally useful to doctors and nutrition advisers for treating people of all ages. It has been established that pollens contain large amounts of goodness including vitamins A, B 12, D, E, C (ascorbic acid) and K. Also present is inositol, biotin, thiamin, riboflavin, nicotinic acid, folic acid and pantothenic acid. Rutin is also available, the richest supply of this essential substance being found in buckwheat pollen, this being understandable when it is realized that rutin is derived from buckwheat. In addition to all this goodness, pollens also contain lecithin, amines, nuclein, guanine, xanthine, hydrocarbons, sterols and growth factors. To the expert in these matters or to the ordinary person it will be appreciated that there is a great deal of goodness in these minute grains. This is only the beginning of the story of pollen. The life force of these grains is without doubt closely related to the high protein, vitamin, mineral, trace element and amino acid content.

Mineral Content

According to the variety of pollen obtained, the ash content may vary from 1 to 7 per cent, giving a general average of 2.7 per cent. These important minerals include calcium, copper, iron, magnesium, manganese, phosphorus, potassium, silicon, sodium, sulphur and titanium. Mineral and trace elements such as those found in pollen were at one time thought to be present in too small quantities to make an impact on the body, although being vital for health. Experts have more recently confirmed that we need only small amounts of these vital substances, especially on the basis of a daily intake.

Amino Acids and Protein

The amino acids in whole dry pollen fluctuate between 10 and 13 per cent and this equals from five to seven times the amino acids in equal weights of beef, eggs or cheese. Admittedly we would not consume anything like the quantity of pollen as we would some of these other foods. Nevertheless we can see what wonderful food value pollen really is, as some of these secrets are revealed. Amino acids are absolutely vital for good health and so play an important part in our lives. There are twenty-one known amino acids and the human body is able to produce all of them except eight, which are sometimes referred to as the essential amino acids. Pollen contains all of the eight essential amino acids and a diet containing quantities of food with these is essential for lasting health. Other foods, known as complete protein foods which have all the essential amino acids include soya beans, wheat germ, liver, meats, skimmed milk, eggs and nuts.

All pollens contain exactly the same number of amino acids, but different species of pollen have varying amounts of each. Whereas some may be packed with certain amino acids, others may have little of these, but larger amounts of others. Such variations depend upon the source of flowers and blooms supplying the pollen and the type of soil where the plants are growing.

Enzymes

A number of enzymes have been found in pollen and these include: amylase, catalase, dehydrogenase, diaphorase, diastase, cozymase, cytochrome, pectase, phosphatase, sucrase, also lactic acids. Pollen compares very favourably to yeast as a complete food when it is revealed that a mixture of fresh pollen can contain up to 1,000 micrograms of cozymase per gram. The alcoholic fermentation of pollen and yeast is identical.

Sugars, Carbohydrates and Fats

There are quite a variety of sugars and carbohydrates in whole pollens in the form of cane sugar (sucrose), fruit sugar (cellulose), grape sugar (glucose). Pollen also has gums, fats and oils. The total fats and oils are only about 5 per cent on an average.

Colour

Pollens vary in colours and shades to the same extent as honeys. The colour of the honey or pollen is determined and controlled by the mineral and other content of the soil and the species of blooms visited by the bees.

Pollen Content of Honey

It has been possible for many years to establish the botanical origin of honey from the pollen content. This can be appreciated when we understand that pollen to plants is as fingerprints to men – a certain infallible means of identity.

Pollen is a highly concentrated substance which never contains above 18 per cent water and some 10 per cent of the total gathered by bees in any particular hive may be removed with no ill effect on the colony. Pollens may contain up to 35 per cent protein, with 15 to 25 per cent amino acids and up to 40 per cent carbohydrates or glucides. Two French doctors discovered that pollen has a growth factor and an antibiotic.

Bee-keepers are increasingly becoming aware of pollen as a health food and there are growing signs that it is being marketed in many countries for this purpose. So in this natural flour-like substance we have a fine food and medicine providing it can be obtained and stored without loss of vitality. It would not pay to try and collect it by employing human labour.

The Cernelle Company has long since discontinued the practice of gathering pollen by means of bees and pollen traps. Instead, large amounts of pollen are gathered direct from the flowers by electro-mechanical machines.

CHAPTER TWO

POLLEN THROUGHOUT THE AGES

To be told that pollen grains may be as old as the earth might seem hardly feasible yet this is possible. In recent years considerable research has been made into this subject, in many countries, and amazing findings have been made from such investigations. Pollen has been referred to as being the fingerprints of certain groups of plants, flowers or vegetables and this is a very apt description. Grains of pollen are like very fine flour but yellowish in colour; each one being so minute that it can only be examined under a microscope. Nevertheless, every single grain is a miraculous masterpiece of intricate construction, having two layer walls, the outer being tough and thick compared with the inner one which is thinner. An analysis of pollen grains can reveal the identity of the plants, flowers, tree or other blossoms from where they originated, for each group has its own characteristics and shape. What is equally remarkable is that the identity of pollen grains, for instance in honey, can establish the country of origin, such as Australia or some tropical land.

Everlasting Pollen

Pollen grains under certain conditions are virtually indestructible, thus paving the way for the survival of man and animals through drought, famine, wars, earthquakes or whatever terrible mishap may befall many unfortunate people on this earth. A methodical investigation of pollen grains will, slowly but surely, reveal changes that have taken place in the plant and animal world since the ice age, like the unfolding of ancient history before our eyes. Much research into this interesting subject has been made. Lennart von Post, a Scandinavian geologist, was one of the foremost experts who revealed many of the secrets of pollen. He proved that pollen grains are so intricately designed that under anaerobic conditions they become virtually everlasting and totally immune from decay. This has been proved when grains have been recovered from deposits millions of years old. From such grains the flowers, plants, trees and so on can be identified; this in turn reveals the history of the plant and animal world. The ability of pollen grains to survive for countless generations is therefore ideal for determining the history of forests.

Pollen Analysis

This subject, termed palynology, deals with the investigation of substances in accordance with their pollen grains. From these findings allergies, caused by pollen in the atmosphere, can be investigated. All growing plants from the smallest weed to the mighty oak tree shed millions of grains of pollen. When it is revealed that it takes 10,000 grains of pollen to cover a surface the size of a sixpence and tons of it fall annually in many countries we get some idea of the vastness of this subject. The amount of pollen in the world is so enormous that the grains of it must be as numerous as sand on the shores. With their inbuilt resistance to decay, grains accumulate over the centuries especially on peat deposits, lakes or bogs.

Throughout the Ages

In many ways our present civilization has much to learn from ancient times. The use and value of pollen is an example of this for it is mentioned in the ancient writings of Egypt, China, ancient Greece, Russia and Persia, to name a few sources. In common with honey, to which it is so closely related, pollen is mentioned in the Bible, Koran and Talmud. Famous and learned men of old, such as Hippocrates, Pliny and Virgil firmly believed that pollen had a most important role to play in ensuring good health and warding off and keeping at bay many of the problems of old age. Such afflictions are endured and taken for granted as being the inevitable lot of man by many people today, even in civilized countries. At one time it was the practice of Hawaiian maidens to collect flowers loaded with pollen from the hala plant during early spring. It was felt by these beauties, and no doubt substantiated, that the pollen in such blooms would enable them to be successful with the young men who appealed to them most. That these beautiful girls still continue to garland themselves with pollen-filled flowers adds strength to this ancient belief.

Egypt and China

Arabs in Egypt and surrounding countries have for generations made use of pollen from date trees as a sustaining food ideal for undergoing hardships encountered in the desert. Likewise, the Chinese have always made use of pollen, combined with honey in the form of cakes. Layers of pollen and honey are kneaded together and spread out to dry when they reach the right consistency. These slabs are cut up into short pieces and after being thoroughly dried are ready for storing. This food, used by many as their daily fare, allowed people to survive during famines, crop failures, or during monsoons. Hunters took such food with them on their expeditions.

The old method of obtaining supplies of pollen, still in force today in some places, may sound very primitive to us when compared to the ideal system of using electro-mechanical devices to gather grains of pollen direct from plantations. It consists of using specially constructed fibre cloth nets to fish masses of pollen from lakes and waterways. After the water is wrung out the pollen is left to dry and all foreign bodies are carefully removed. When perfectly dry, the pollen is crushed up into powder form and stored in clay containers which are sealed. Apart from food, the descendants of these ancient people use pollen as a medicine and antiseptic. We now know such methods can present problems owing to there being impurities in the pollen so obtained.

Russia

The waste matter which collected in the bottoms of hives was used by old time bee-keepers in Russia and many other countries. Many of them were too poor to use the honey, which they were compelled to sell. Poverty often proved a blessing in disguise for these people. Necessity forced them to dispose of the bulk of their honey supplies, so they consumed the bee scrap from the bottom of the hives. Perhaps they did not really know the true food value of this scrap.

The substance treated as honey scrap has long since been shown to be one of the finest foods in the world, being almost pure pollen. During the earliest days of the Olympic Games the competitors ate natural honey to enable them to have energy to give of their best. As with the bee scrap, this honey, being unstrained, contained a large amount of wonderful pollen. Perhaps these ancient country folk were immune from any impurities which might have appeared in pollen obtained in this manner.

Pollen in the 80s

Since about 1950 pollen has been used as a food supplement and today research into Swedish Cernitin pollen is being carried out in some twenty countries. During the past decade more pollen products have appeared on the market and a large number of pollen extracts as health food supplements and as an aid to overcoming many complaints are exported from Sweden, Austria, France and other parts where pollen processing is taking place.

One Swedish pollen extract has the ability to eliminate the toxic effects of bacteriotoxin streptolysin which, if left unchecked, can destroy body cells, causing weakness and loss of resistance, which may well result in disease. This can be observed when the substance is put in a test tube of blood and the red corpuscles are destroyed; but, if the pollen extract is also added at the same time, the bacteria toxin is gradually neutralized.

For some years a Cernitin pollen based wound ointment has been shown to hasten the healing of abrasions, wounds and other similar conditions.

Perhaps the most outstanding success with Cernitin pollen is with Pollitabs Sport, which have been used with excellent results by ice hockey teams, long distance runners, canoeists, rowers, wrestlers, weight lifters, footballers, and the champion boxer Mohammed Ali.

CHAPTER THREE

SWEDEN: HOME OF THE POLLEN INDUSTRY

Until the early 70s pollen as a food supplement was almost unheard of in the United Kingdom; but nowadays it can be obtained all over the country in many different forms, mostly from Health Food Shops. In spite of this the average person knows little about putting pollen to everyday use, especially athletes and those engaged on arduous tasks. Prior to 1970 it would have been virtually impossible to gather pollen by the ton but for a long time now this stage has been reached by the Swedish company Cernelle, with its modern ways of harvesting.

Costa Carlsson – Pioneer of Pollen

In order to understand how this success has been achieved we have to go back some 30 years to the work of a Swedish amateur beekeeper and railway employee, Costa Carlsson. After much research he astounded beekeepers, and others, by his invention of the 'Pollen Food Harvester', forerunner of the present-day sophisticated equipment used by Cernelle to extract pollen from plants without resorting to the aid of bees. Like beekeepers and others devoted to the soil, plants or animals, Carlsson found a great deal to interest him in his hobby. For some ten years he made attempts and finally achieved what others thought to be impossible; getting pollen direct from the flowers without robbing the bees of their essential food.

Food for bees consists of water, honey and pollen. Honey supplies the energy and pollen contains all other essential substances, such as fat, vitamins, minerals, sterines, hormones, enzymes and co-enzymes. Therefore, pollen is essential for the survival of the bees in the hive. If too much is removed from any particular hive the unfortunate colony would surely die. There is no substitute that man has devised to replace pollen, for it is the bee's protein, body and life. This is one of the vital differences between pollen and honey. In most countries with cooler climates such as England, the entire honey content is extracted from beehives; the colony being fed during the winter months on a sugar-like substitute on which they can survive without any apparent ill effects. In spite of this, the author prefers to obtain honey from Australia, where the crop is so large that sufficient for the need of the bees can be left in the hives all the year round. As a result the Australians, with their wonderful climate, can guarantee to supply honey from, 'non sugar-fed bees'.

Before Carlsson's invention, a classic pollen trap was in use. It consisted of a series of mesh-like enclosures through which all the bees had to pass on entering the hives. In so doing the harvester caused the raw pollen pellets to fall from between the bee's legs into a special container below.

From these early beginnings a massive pollen market has been built up in Sweden and the pollen harvester has been superseded by a series of machines based on electromechanical principles. This enables the company to gather quite large harvests of pollen direct from plantations of flowers in a vacuum-like process without the aid of the bees. From this early beginning a large pollen market has been developed in Sweden with a head office at AB Cernelle, Vegeholm. The research work was carried out during the early years at AB Kabi, Stockholm, where a world-famous bio-chemical factory has been built. So this patented device proved a turning point, not only in the life of Carlsson, but more importantly, in the world of pollen. Before this invention there was just not enough pollen available for scientists and medical men to be able to make their astounding discoveries about it. Carlsson changed all this by ensuring that sufficient supplies could be obtained for full scientific investigations.

Two Decades of Research

By his ingenious means of getting supplies of pollen and his subsequent study of bees, Costa Carlsson made the Cernelle company of Sweden world famous. The Pollen Food Harvester was only the beginning, but nevertheless a vital link making possible the study of large quantities of pollen for the first time. It had been known for generations that a grain of pollen, minute as it is, holds all the elements of plant life, vitamins, minerals, hormones and nuclei acids. Like honey and royal jelly, pollen contains certain elements which still defy identification. It was a long time before man was able to find a way of piercing these tiny grains. Pollen has baffled us for years and it will take much more research before all the secrets are known. To unravel many of them the firm of Cernelle built a laboratory equipped with the most up-to-date scientific instruments. The site of this building was carefully selected in an area surrounded by field and pine trees in Southern Sweden. From here the story of pollen is being gradually revealed and other ingredients identified. So this interesting programme continues with more scientific experts and sophisticated instruments.

Palynology

The study of pollen, known as the science of palynology, has only really been established on a large scale during the last three decades, mainly in Sweden. The palynological laboratory in Sweden, possibly the only one of its kind in the world, has done much to bring pollen within the reach of everyone. During the course of the study of pollen, experts refer to it as the fingerprints of plants, trees, flowers, weeds or vegetation. This is more clearly understood when it is realized that pollen from each and every single bloom bears characteristics relating solely to the species from where it originated. Methods of identity concern the size, shape, weight, ridges, hollows, and number of germinal openings (known as hila) on each and every minute grain. Swedish experts Lennart von Porat and Gunnar Erdman were pioneers in the field of pollen investigation. All persons who have benefited, all over the world, from this wonderful food and medicine owe much to the work of these two scientists.

Method

The many experts engaged in these fascinating investigations of pollen are termed palynologists. First they obtain a sample of the soil, which is subjected to treatment with potent acids – whereupon everything vanishes, dissolved by the acids, or we should say nearly everything, for the tough and indestructible grains of pollen remain. This fact more than amply demonstrates just how remarkable these minute, but mighty, grains of pollen really are for their sheaths will survive when all else perishes. From a detailed examination of the grains recovered the identity of the various species of vegetation can be ascertained. When archaeological excavations are being made a detailed investigation of the pollen grains will build up a complete picture of the type of vegetation growing at that particular time. From this much can be learnt.

Cernitin Pollen

With the tremendous breakthrough of being able to unlock the mighty pollen grains, Carlsson and his team of scientists were able to obtain supplies of pollen nuclei. Cernitin is in fact a group of pollen extracts from organically grown, specially selected flowers. Each drop of cernitin is packed with the nucleus of millions of pollen grains. Cernitin also contains twenty-one amino acids, all the natural vitamins, minerals, hormones and nucleic acids. This precious pollen nuclei has other unknown ingredients, essential to plant life. Expert knowledge and their means of identifying of grains enable palynologists of the Cernelle Company to segregate the various kinds of pollen. To illustrate how accurate they are, it is interesting to note that of the large amount of raw pollen collected, the total foreign grains extracted never exceeds 1 per cent. So the story of pollen continues and further progress is being made.

Pollen Tablets

The Cernelle Company distribute pollen tablets throughout many countries in Europe and overseas. This product contains cernitin, the organic extract of pollen, completely pure and unadulterated. No synthetic substances are used and there are no hidden dangers or the side-effects sometimes present in drugs. What is perhaps most important is that persons allergic to pollen can take even large amounts with no adverse effects whatsoever. This is because, as stated, pollen gathered via the bee has been treated by this clever creature with nectar to overcome the cause of allergy. Only airborne pollens, in a raw state direct from trees, vegetables and flowers present the dangers of allergies. These tablets have proved especially good in warding off the common cold and each one contains 30 mg of cernitin.

Cernelle Special

These are tablets which contain 60 mg of cernitin and in addition to minerals and amino acids, each has the following vitamins:

Vitamin B1	1 mg.	Vitamin B2	1 mg.
Vitamin B6	1 mg.	Vitamin C25	25 mg.
P-P factor	7 mg.		
Calcium pantothen	3 mg.		

Thirty Years' Success

How safe, good, and widely accepted are pollen tablets? Well, the record of the Cernelle Company of Sweden is most impressive. Cernitin pollens in various forms have been sold to the public for three decades with sales increasing steadily all the time. When it is realized that at present over 400 million cernitin tablets are sold every year some idea is gained of the large number of people who are satisfied customers. This is quite an achievement, but what is most interesting to potential customers is that, in spite of this large turnover, not a single case of any side-effects or allergies have been reported in the use of pollen. This evidence is very important and provides definite proof that natural medicines and foods are best, most health giving, and safest in the long run, especially when we consider the side-effects and dangers sometimes present in the use of drugs.

Treatment of Many Illnesses

Doctors from many hospitals have reported success in the treatment of various complaints, which they attribute to the use of cernitin. Such illnesses include influenza, urinary disorders, and measles.

All this is really not so surprising when it is realized that long before the advent of the Swedish product, pollen was frequently used by doctors in many countries to overcome such allergies as hay fever. All this account of activities in the world of pollen is really only just the beginning. Extensive tests are still being carried out in many countries, in particular Sweden and the U.S.A.

Pollen by the Ton

Mention has already been made of the experts' opinion that literally tons of pollen appear each year in every country of the world. This is nature's way of ensuring that species of tree, plants and vegetables remain with us for future generations. The greatest amount of this pollen disappears into lakes and rivers, but some finds a place in bogs or marshes where it will remain for generations on end – just like seams of coal which may have been deposited in England and elsewhere in the Stone Age and are still with us today.

The total amount of raw pollen harvested, prepared and marketed by the Cernelle company depends on their existing stocks. The record annual tonnage handled by them was 40 tons. This is a tremendous undertaking indeed when we consider the billions of grains it takes to make a ton of pollen and the work involved in collecting and preparing it. Approximately 400 to 500 million tablets are produced annually.

World Wide Market

Cernelle export their various kind of pollen in tablets or other forms and their world famous beauty preparations to no less than fifty-four other countries. Among this vast number of lands that import pollen as a dietary food supplement or natural health product, no less than 16 recognize pollen as a pharmaceutical speciality; these include Argentine, Austria, France, Germany, Greece, Japan, Spain, Switzerland and Russia. The large number of pollen preparations marketed by Cernelle include some animal health products. I quote these for the benefit of animal lovers, pet owners, farmers, and so on. They are: LBC Lactic Bacteria Cernelle, Cavallifex, Cernidog, Cernisex and Cernitin Dog Ointment. It is only a question of time before pollen in its various forms as a health food, medicine or aid to beauty will be available in every country of the world.

Latest Developments

Carlsson's original pollen trap has now been superseded by a series of machines based on an electro-mechanical principle. This new method enables Cernelle to collect pollen direct from flowers without any aid from the bee. The company at present markets no less than forty-three different pollen products.

Raw Pollen

According to the Cernelle company, the pollens gathered by bees for their store in the hives could perhaps present a health hazard to humans if consumed in its raw state. Such pollen is often contaminated by such foreign bodies as bacteria, fungi, mites and insect eggs, which may well produce an antigenic effect.

After decades of study into methods of obtaining pollen, the Cernelle company is continually improving its systems to ensure the end product is entirely safe and free from any substance which might cause problems to consumers. One way in which this is achieved is by obtaining supplies from specially selected flowers which have not been contaminated by insecticides or pesticides.

CHAPTER FOUR

POLLEN IN ENGLAND

The climate of England is really far from ideal for breeding large numbers of bees compared with Mediterranean countries and such distant lands as Australia. So England is not a country where large supplies of either honey or pollen can be gathered. No doubt for this reason English honey is more than twice the cost of that from Australia, in spite of the added freight charges for a 12,000 mile sea voyage. Should English bee-keepers extract pollen from their hives on a large scale the same higher costs would no doubt apply. Fortunately, as with honey, we are able to get supplies of pollen from other countries, especially Sweden, the world's experts in this field.

As shown in Chapter Three, it was in Sweden that the first and greatest large scale development in the world of pollen and in the utilizing of it as a food and medicine took place. In England we are becoming increasingly aware of this fact, in particular the progress being made by the famous Swedish firm of Cernelle. Cernelle are experts in the cultivation of every type of pollen which they are able to harvest by the ton, an undertaking which would have been considered impossible. As a result of the progress made in gathering pollen they are able to export it all over the world.

During 1967 Alan Randerson, then Managing Director of a soft drinks company, visited Sweden on business. His first introduction to pollen was when little tablets were served up on the breakfast table by his Swedish friends. As events will show it proved a turning point in his life and career for he immediately became fascinated in the nutritive value of pollen and made an extensive study of this absorbing subject. He was amazed at the wide use of pollen which had been taking place in Sweden and other countries for a number of years, while in England it was practically unheard of in medical and health food circles.

Doctors in Sweden widely accepted pollen and over 4,000 of them prescribed it as both a food and medicine in treating patients suffering from varied complaints. As Mr Randerson's knowledge of Swedish pollen grew he realized how important it had proved to be for retaining good health and in curing persons of numerous maladies. This eventually culminated in the formation of a company producing pollen in London, in 1968, of which this enterprising gentleman became the managing director. An apt title often bestowed upon Mr Randerson for his efforts in introducing pollen to this country is 'Pollen King of England'.

Why Swedish Pollen?

Pollen is the same all the world over whether obtained from cultivations in Sweden or elsewhere, providing the source is forestry, or vegetation under natural conditions or if planted by man in a chemically free environment. The world market is controlled by Sweden because they are experts in seeking or planting the best possible cultivations and lead the world in the art of collecting and processing pollen. It is for these reasons that Mr Randerson obtains his supplies of pollen from Sweden and as an added safeguard they come from a laboratory which is under the strict control of the Swedish Royal Medical Board. A very important problem to be overcome when supplies of raw pollen are collected for human consumption is that the raw supply contains impurities and so needs expert cleansing, purifying and processing. This laboratory was the only one which could guarantee expert processing of pure pollen, for the company was determined to use pure pollen and not an extract.

Pollen Tablets

There are a variety of guaranteed pure pollen tablets which contain this health giving substance in its most natural form, as distinct from the extract of pollen which is sometimes produced. There are two kinds of pure tablets, one fortified with vitamins, B1, B2, B6 and E, made easy to take by being sweet coated. The other variety contains a higher intake of vitamin C. Supplies of these tablets may be purchased from Health Food Shops.

Pollen and Honey

A health food supplement, consisting of pure pollen and honey, was introduced some ten years ago. Even the hard working little bee has never had the good fortune to be provided with these two health giving foods blended together, he has had to do this for himself. Such a wonderful combination of goodness has not been provided from the beehives since royal jelly was discovered. Naturally, pollen and honey costs more than pure honey alone but this is offset by the fact that the amount needed daily is as little as half a teaspoonful.

Export Market

In spite of being a comparatively new venture, pollen production is expanding rapidly. One company now exports pollen to the following countries in addition to the home market: Australia, Hong Kong, New Zealand, South Africa, North and South America, and Holland. If this upward trend continues they will soon be exporting more than they sell in England. Certain amounts of pollen are sold through the retail trade in England but the bulk of the sales are to individual customers by mail, mostly through a monthly delivery. Business has often become so brisk that supplies have run out, particularly the new pollen and honey concentrate, so increased production has been necessary. The future looks very bright for pollen as a health food in England and many overseas countries.

Beauty from Nature

Progress is being made with pollen cosmetics in England and this side of the business is also expanding. Pollinated night cream is available containing pure Swedish pollen with added wheat germ oil which is rich in vitamin E. This night cream has proved to be a wonderful skin cleanser, which, if given a fair trial will greatly improve the skin in every possible manner. Cosmetics made from pollen, being natural, have no harmful side-effects.

CHAPTER FIVE

THE WORLD OF THE BEE

Almost without exception every flower or fruit blossom produces a sugary substance known as nectar at the base of its petals. After being gathered by the bees and converted into honey, this nectar has been acclaimed as one of the finest foods for man from prehistoric times.

Honey and Bees

Bees, bumble bees, butterflies and some other insects all have one thing in common: that amazingly designed proboscis which unfolds from their mouths. With this they are able to draw nectar into their stomachs from every variety of flower or blossom. Unlike our insides, bees have a gut designed as a storage compartment, honey sac or shopping basket. So each bee is able to carry the contents of the honey sac back to the hive, for storage in the cells. By means of a valve the bee's stomach also serves the purpose of supplying the creature with sufficient food. Hunger is satisfied by releasing the valve till enough nectar enters that part of the stomach used for sustenance; but only nourishment allowed past the valve is consumed by the bee.

The bee is able to perform the remarkable feat of flying back to the hive, carrying his own weight in pollen or honey, or a mixture of both. This is especially outstanding when it is considered that even the most modern aircraft can only be loaded with up to 25 per cent of its own weight in order to facilitate a safe take-off and landing.

Bees and Pollen

Without pollen there would eventually be no plants, trees or flowers of any kind – it would be similar to the human race becoming impotent. We would simply die off, so would all plants if deprived of pollen. So it will be appreciated that these tiny grains of pollen contain a mighty power. Just as the human body needs lots of protein to develop from the baby stage to manhood, so it is with the bee larvae, brood – and particularly the very young bees. Throughout the lives of humans a constant supply of protein is vital to repair daily wear and tear and breakdown of muscle fibres. The bee has access to a constant supply of nectar to provide energy needed to carry out the gigantic task of daily foraging. Remember, it has to visit some 1,500 or more blooms before filling the honey sac. It is the gathered pollen, very rich in protein, that supplies this little creature's daily intake in addition to filling a store in the cells of the hives. Unlike us, the bee must consider the winter months when blossoms are no longer available in many lands and temperatures are too low for the bee to work. So in every hive the bees gather large supplies of pollen in the spring and summer. Pollen, after being mixed with honey in the hive, is carefully stored away to ensure survival of the colony throughout the most severe winter. So to deprive a bee colony of pollen would be akin to cutting off the protein supply for us, our bodies would gradually deteriorate, so would the bees'.

The bee's body is constructed in such a way that he simply is unable to help collecting pollen grains on it: even if he wanted to return to the hive without pollen, this would be well nigh impossible. As a bee enters a flower to syphon nectar the minute pollen grains cling to his wonderful hairy legs specially designed for this purpose. Only airborne pollens bring allergies, for bees on gathering pollen add nectar and saliva which ensures the grains are completely safe.

Bee-Bread

The pollen which bees bring to the hive (already combined with nectar) is then mixed with honey and stored in the comb cells, where, after lactic fermentation, it becomes 'bee-bread'. On bee larvae being fed pollen or bee-bread their weight increases no less than 1,500 times in less than one week. This fact shows what marvellous food pollen is for body building, as far as the bee is concerned at any rate. Bee-bread has all the ingredients of pollen plus vitamins E and K. Ordinary pollen commences to deteriorate very gradually on storing but bee-bread resembles fresh pollen even after two years and retains its food value.

No Pollen – no Bees

The role played by pollen in our lives is probably so complex that it will take man a long time to fully explore this most fascinating subject. Intensive study of bees in many countries has proved that these energetic little creatures are unable to live without pollen. No wonder they devote equally as much time to the collection of pollen as they do to nectar which they store as honey. Worker bees continually gather pollen to preserve in the hives as bee-bread. What a daily slice of bread it produces for these magnificent creatures, especially for the queen, for it contains all the nutrition needed to feed her for a period up to five years! Her life spell would be very limited without it. In the hive the bee larvae lives on pollen, and as young bees appear from the larvae, they too exist on pollen for their first food. We can see how nature has made up this miraculous cycle, for without pollen not only would plants die off but the bee's doom would also be sealed.

The Beehive

The hive must be the centre of all discussions on honey, royal jelly and pollen. There has always been a certain amount of mystery surrounding the little bee and his hive. Many questions have been answered but some may always remain a secret, shared only by the bees. For centuries it was always noted that, generally, bee-keepers tended to live long active lives. Studies of the life of the bee shows that the queens become super creatures in every way, in spite of beginning life in exactly the same way as any other member of the hive. This study is a most fascinating subject and Professor Dyce, Cornell University in America, established that in the larvae stage the queen and worker bees are absolutely identical. So life for the queen begins when she is hatched from an egg similar in every respect to the other millions of eggs which appear in the hive. She is born equal to all other inmates.

Eggs Galore

What is perhaps the most extraordinary thing is that during the entire life of the queen she is able to lay up to 2,000 or more eggs every day. This is quite an astonishing feat but all the more remarkable when it is confirmed that such a daily batch of eggs equals more than twice the weight of the average queen, who, it is estimated, will produce over 2,000,000 eggs during the course of her life. In spite of such achievements, intensive study of the hive for many years has confirmed, beyond doubt, that the only difference between the queen and the ordinary worker is her diet. This is a classic example that even in beehives tiny creatures, like human beings, simply become what they eat. The nursing bees prepare partly digested pollen which they mix with a chemical from glands secreted in the top of their heads: this is known as 'bee-bread' or pollen mush. This food is so sustaining and abounding in vital goodness that within two days of feeding the larvae, tiny bees appear. It is at this point that the life of the tiny bee selected to become queen differs from that of all other bees in the hive. The ordinary bees switch to a diet of honey but the queen, upon which the entire hive depends for survival, stays on a diet of bee-bread all her life.

Bees-wax

Bees-wax was always thought to be made entirely of pollen. This was disproved by Swiss naturalist Francis Huber in 1936. In his book *New Observations upon Bees,* Huber describes the following interesting experiment.

He placed a swarm of bees in a room, without a comb, but supplied them with honey and water. To his surprise, within five days the bees had constructed a number of combs. When these combs were removed the swarm immediately set to work to build more and this continued until seven batches of combs had been built. Huber proved that bees-wax can be produced by bees from honey, plus whatever these insects secrete to make the wax. However, this is only carried out by a swarm in an emergency: if deprived of pollen for any length of time, the bees will perish ultimately.

Life Force

That a swarm of bees cannot live without pollen was amply demonstrated when the experiment of comb building was continued for a longer period. Lack of pollen caused the unfortunate swarm to die off from utter exhaustion as a result of building the combs. So if a swarm of bees do not have access to pollen, their life force, they will perish fairly quickly if, at the same time, they are called upon to build combs of bees-wax.

From his study of bees Baron von Berlepsch was able to show that a swarm with no pollen will utilize nineteen pounds of honey to produce one pound of bees-wax. If the same swarm has access to pollen the amount of honey used is reduced to twelve pounds. So pollen is essential for bees to produce large amounts of bees-wax. Once again this vital substance is shown to exert a mighty power in the world of the bee.

Three Wonderful Foods

Since the first bees appeared on earth, honey has provided man with one of his finest foods and this fact is indisputable. In recent years the wonders of royal jelly have been pronounced, a jelly-like honey claimed to possess fantastic powers and the sole food of the queen bee. For human beings it was reputed that royal jelly would act as a revitalizer, giving strength and energy to tired bodies. The writer, an ardent gardener and nature lover, has always appreciated the tremendous work carried out by such friendly creatures as the earthworm, ladybird and bee. However, in spite of consuming honey in place of sugar for over fifteen years, I considered the claims made about the wonders of royal jelly to be exaggerated, and my belief was that the bees were the only ones who could support this claim for their jelly. Bees unfortunately are unable to talk.

During the bitter winter of 1963 after the late President Kennedy started such a scheme for members of the Senate, the *Daily Sketch* organized a 'Pace Setters' walk of over fifty miles from London to Brighton. The writer entered and completed this walk in some fourteen hours. Many gave up owing to the intense cold of the bitter February night and following day. As the walk progressed, competitors expressing a liking for honey were invited to taste samples of royal jelly; those who wished were also handed jars of this substance to help them on their long walk. I, with many others, accepted this kind offer but with considerable reservations as to the mighty claims made by the proprietors about their jelly. However, in the years that have passed since, my opinion has gradually changed after intensive study of the bee and his way of life. In particular, now that the vitamin, mineral and other content of royal jelly has been revealed, it seems there are good reasons for some if not all of the claims made for this food from the hive. This feeling has been strengthened because, on occasions, having sampled the jelly during times of stress or very violent exertion it has enabled me to carry out tasks which seemed beyond my capabilities. In more recent years the feeling persists that royal jelly did more to help me complete the long walk than I realized at the time. (The exercise proved comparatively easy at the age of forty-six, in spite of the sub-normal temperature in that bitter winter.) As with many other things in life, people will need convincing that pollen is good for them. So with pollen, or royal jelly, the answer is to obtain some and try it out for yourself for the sake of the small sum of money to be expended. The writer cannot endorse the opinion of some people that to consume a jar of honey will solve all our matrimonial or other problems within days. However, long years of experience has proved to me that honey is wonderful for our health, royal jelly is even better, but pollen, holding the key to everything in the beehive, is superior to both. All the evidence available supports the belief that pollen is the most mighty find from the world of the bee.

CHAPTER SIX

POLLEN AS A FOOD

Pollen is a natural food guaranteed pure and unadulterated, being free from insecticides. As a supplementary food, the daily intake of pollen can be about 20 grams for persons enjoying good health. However, for those in a poor state of health, invalids or elderly persons who need a total revitalization, from one to three tablespoonfuls may be taken daily.

Pollen Taste

There are as many varieties, colours and varied tastes in pollen as in honeys. Many people on sampling it for the first time are very disappointed, being inclined to think that because it is closely related to honey it will have the same delicious taste. Unfortunately this is often just not so, for invariably pollens are somewhat bitter. Exceptions to this rule are the sweet ones, which provide a more pleasing taste. So do not be put off this fine food on taking it for the first time owing to this bitterness. Many of our health supplements are not as pleasant as honey to take, a typical example being brewer's yeast, one of the finest foods in the world. Yeast tablets are packed with the B complex vitamins but are somewhat bitter, like pollen. This can easily be overcome by swallowing the yeast or pollen tablets. To make pollen more appetizing it is simply blended with dates, honey or fruits. An example of this is the pollen and honey preparation described in Chapter Four. The two substances are mixed in a liquid form to give a taste similar to less sweet honey.

Tests

In 1957 a French doctor named Chauvin announced to medical circles the results of his extensive study of pollen as a food for human beings. Almost two tons of pollen was used directly as a food supplement by a number of experts in medicine and nutrition. All groups of people were involved, from small children to elderly persons. The results of this intensive study showed pollen to rate highest on the list of nutritive foods.

This doctor showed pollen to be excellent for the cure of anaemia, especially in very young children, for it quickly gave them an outstanding increase in red blood cells. Tired bodies were rapidly restored and in particular it had a revitalizing effect on the old, giving them a new lease of life. It was effective in curing chronic constipation, flatulence and infections of the colon, especially diarrhoea. After severe illness or shock it was found that pollen quickly restored health and strength: a return to normal weight was one of the most wonderful things repeatedly noticed about those underdeveloped persons given this fine food. For centuries the great food value of honey was accepted but in more recent years royal jelly was shown to be even better. Now pollen proves to be the most health giving strength and energy restorer of all foods from the beehive. What proves most gratifying about pollen is that it is not only non-addictive but shows not the slightest ill or side effects from persons taking it, even over long periods.

Perfect Food

Pollen has shown itself to be a complete nourishment in every sense of the word. It would prove hard, almost impossible in fact, to find a food from animal or vegetable sources containing such vital nutritional elements. Not only does it build up strength and energy in tired bodies, but it acts as a tonic. For people who have temporarily lost the zest for living, a course of pollen may well be the answer to their problems. French doctors have noted that in less than a week in practically every case pollen restores normal healthy appetites to people who previously have seldom enjoyed meals. This is perhaps why in most cases it increases the bodyweight of under-developed persons. Strangely enough, pollen was also found to aid weight reduction for people who were fat, proving that pollen is an ideal body regulator in every possible way. This new food was shown to stimulate many functions of the body including the gastric system. Containing a natural antibiotic, it also controls dangerous bacteria in the intestines. The benefits experienced by persons given pollen by French doctors not only consisted of restored bodily health, but also a more optimistic outlook on life, which is so often closely related to physical harmony.

Weight Gains

After taking pollen, underweight persons have invariably made lasting gains and a general all-round improvement in health, increased appetite and feeling of well-being follows. People have more vigour, vitality and increased resistance to infection.

Athletes

Some years ago American footballers proved pollen to be an ideal body building food. Tests were conducted for fifteen weeks with two groups, one receiving pollen tablets daily and the other a multi-vitamin supplement. At the end of the period those players taking pollen showed an average five and a half pounds weight gain, whereas the vitamin group's average weight remained static. Trainers consider that footballers, almost without exception, lose weight after a season's play. Vitamins proved that they could retain their weight but pollen did more by providing an actual increase: not simply a gain in fat but protein from pollen, a gift from nature carried by her marvellous army of bees. This provides proof of the body building potential of pollen for humans even under extreme conditions of violent exertion, when one can, at the best, expect to lose weight.

Enormous Vitamin Content

Various experiments concerning vitamins were conducted at the American Universities of Minnesota and Wisconsin in 1942/43. From the latter university the findings from a analysis of pollen and royal jelly for B group components confirmed a fluctuation of from 49 to 378 times as much B complex vitamins in pollen as in honey. Concerning royal jelly, the amounts varied from 101 to the fantastic figure of 6,212 times the B group vitamin found in honey. There are of course many reasons for this great variation, for as previously explained, the vitamin content of honey changes considerably in accordance with different plants, soil conditions, age of the honey, amount of pollen present and so on.

Pollen in Honeys

The amount of pollen appearing in honey can be controlled by the ways of processing and the type of nectar. Nectar from some plants may have only a low pollen content but from others the amount can be very high. Pollen also varies in relation to the type and size of blooms and arrangement of anthers. During honey processing, pollen may appear on the surface like cream on milk.

Pollen Supplements

Pollen in various forms may be taken daily as a food supplement for any period of time. It is considered best to take the pollen capsules, or pollen combined with honey, on an empty stomach. The ideal time is thought to be between fifteen and twenty minutes before a meal, preferably breakfast, when the stomach is completely empty. It can equally well be taken before other meals during the day: a course lasts one month.

Pollens are graded by vitamin, mineral and amino acid content. Certain pollens may cause allergies, but to repeat what I have already stressed, those pollens collected by bees do not. Bee-gathered pollen may safely be taken by everyone, even those persons prone to allergies such as hay fever, as they will suffer no ill-effects from the pollen. Not only does it have no side effects, but it is non-habit-forming.

A French doctor, Alin Caillais, considered from his vast experience that an average daily intake of pollen of about 20 grams can supply the physical requirements of the average person. A course of pollen may be taken yearly or more frequently according to the need. Pollen, the health food of the future, is easy to obtain, comparatively cheap and has been shown to have lasting benefits. It may safely be taken by persons of all ages, in fact the older people are the more they have to gain from the restorative powers of these tiny grains presented to man by nature.

Pollitabs Sport

A remarkable Cernitin pollen extract, Pollitabs Sport, has been used with excellent results by ice hockey players, long-distance runners, canoeists, wrestlers, weight lifters, footballers, and many other outstanding athletes.

Pollitabs Sport was successfully tested as long ago as 1963 on an ice hockey team from Rogle, Sweden. This team was upgraded from series 4 to series 1 in five years, a record for the sport.

Finnish long-distance runners, not very successful in the 1968 Olympic Games, reformed their diet under the direction of famed New Zealand trainer Arthur Lydiard. On the advice of Lydiard, the runners also took Pollitabs Sport and Polloton pollen extracts, increasing their training distance from 5 to 25 kilometres per day. Subsequently the Finnish team won several gold medals at the Munich (1972) and Montreal (1976) Olympic Games.

CHAPTER SEVEN

POLLEN IN EUROPE

During the early 1950s many countries started what subsequently proved to be a boom in the production of royal jelly. Medical opinions after extensive tests on humans and animals paved the way. France, a large agricultural country, many parts being ideal for bee-keeping, became one of the leading royal jelly producing countries. Four interesting books were written by notable French experts between 1954 and 1960 on the subject of royal jelly and pollen. The titles and authors of these works are as follows:

Le Miracle de la gelee royale. R. Dubois (1954)
Les trois alimento miracles: le miel, le pollen, le gelee royale Alin Caillas (1957)
Le gelee royale des abeilles. B. de Belvefer (1958)
Manuel pratique du producteur de gelee royale. Alin Caillas (1960)

Institute of Bee Culture

Credit must be given to Dr Emil Chauvin, Institute of Bee Culture, Bures-sur-Yvette, France, for research into the food value of pollen. Intensive study of the effects produced by feeding it to thousands of animals showed that there were no adverse effects. In the case of mice, after hundreds of them were given pollen for two years a study of these creatures and their breeds showed them to enjoy perfect health, greater endurance and a higher rate of reproduction. Further tests likewise proved pollen to have an extra high proportion of life-giving properties if employed as a food for human consumption. In the case of all age groups, children, adults and elderly persons, the results were excellent, supporting the belief that it is the greatest discovery ever as a restorative food for building health. It brought vital life-giving strength rapidly to all and for the older persons results were so remarkable as to constitute a 'rebirth'. The great value of pollen was made known in world medical circles by Dr Chauvin in January 1957.

Pollen for Health

Two French doctors, Chauvin and Lenormand, discovered that pollen contains both an antibiotic and a growth factor. This antibiotic, of which penicillin is a prototype, also prevents the growth of some microbes. It therefore regulates the intestines in that it will destroy or weaken harmful bacteria, but at the same time promote the growth of health giving species. This is an important but less spectacular bodily benefit that can be derived from a daily intake of pollen. To quote the findings of these two doctors in their own words, after these extensive tests: 'Pollen acts as a tonic, rapidly restoring normal weight and energy to a convalescent.'

These two doctors also showed that pollen produces an all-round improvement in general health, being ideal for invalids or persons recovering from illness or operations. It makes for a speedy return to health and restoration of normal weight. Dr Chauvin was able to show that pollen will cure chronic constipation when other remedies fail. Likewise it is wonderful for intestinal complaints including diarrhoea, inflamed intestines and other infections of the large intestine.

In January 1957, these doctors reported to members of the Academy of Science and had this to say about pollen:

It has a regularizing effect on the function of the intestines. For anaemic children it produces a quick increase in the haemoglobins in the bloodstream, bringing weight increase and vitality.

They strongly recommended a course of pollen for mentally retarded children or those suffering from stunted growth.

These findings were clearly demonstrated when children in a convalescent home near Paris were given pollen for periods of up to two months. The doctors examining these children noted increases in both the red blood corpuscles and in the amount of haemoglobins, the former by up to 30 per cent and the latter by 15 per cent. These findings clearly showed how good pollen is for combating anaemia.

From their study of pollen or bee-bread, French and German scientists found it to be an ideal addition to the diet and that after taking a course persons could expect a feeling of well being, increased vitality and resistance to illness. Steady nerves and better appetite closely related to improved general health.

Belgium

Belgium is a small country with a climate not ideally suited to large scale bee-keeping, but in spite of this a thriving royal jelly and pollen food business exists. Belgian royal jelly and pollen comes from some of the finest bee-keeping centres in France, which are under direct control of the parent company in Belgium. As a result they are able to supervise packaging of the jelly directly after harvesting. This royal jelly obtained from organic sources is guaranteed pure, wholesome, and will remain in perfect condition for a period of up to eighteen months. It is readily available in England and many other countries from Health Food Stores.

Pollen from Belgium

Success with royal jelly in Belgium encouraged the build-up of a flourishing pollen industry, both in the home market and for export. Belgium products are made of flower pollen, lecithin, seaweed and buttermilk. The last three ingredients have been accepted for a long time as having protective and vital health giving properties. So when, in addition, all the goodness of pollen is added we are certainly getting good value. Pollen is the main and most vital component and it may be taken in water or with any other drink. The other foods included are rated highly for the following reasons.

Soya lecithin, derived from the amazing soya bean, is wonderful for the entire nervous system. Apart from containing iodine, essential for a healthy thyroid, *seaweed* has also some ten mineral salts and many trace elements. *Buttermilk* puts the finish to the contents, making it pleasant to taste and assisting the body to assimilate all these fine foods in one, king of which is, of course, pollen. The conclusion which the Belgian company draws after years of research in France and elsewhere is that pollen, collected by the bee, can, in every sense of the word, be classed as a magic food.

Austria

Austria, like Belgium, has a flourishing pollen food industry which exports products to many overseas markets. Supplies of pollen are obtained from many countries for processing in Austria and their pollen capsules also contain royal jelly and all the other goodness and vitamins of pollen.

A great deal of work has been carried out by Austrian medical experts into the use of pollen for treating all manner of illnesses. The Austrians have done a great deal to prove the value of and promote the use of pollen. A brief summary of some of the work carried out by Austrian doctors is described in Chapter Nine.

CHAPTER EIGHT

POLLEN IN THE UNITED STATES OF AMERICA

In 1960 the annual amount of honey obtained throughout the United States of America was about 80,000 tons. It was estimated that the same amount of pollen could have been collected. Naturally, a market for such a vast amount of pollen must exist to make harvesting and preparation repay the time and expense involved. A great fluctuation exists between the amount of pollen gathered by bees in various hives. Some colonies are unable to collect very large amounts during the entire year. However, it is not unknown for certain hives to collect as much as sixty pounds in just the spring period, although this is unusual. The honey bee is able to transport his own weight in either pollen or nectar and is able to fly at speeds approaching fifteen miles an hour. Bee pollen is in fact a mixture of pollen, honey (nectar) and saliva.

Pollen Traps

Although credit for success in the invention and extensive use of the modern pollen trap undoubtedly goes to Carlsson of Sweden, traps were experimented with a long time before. It was in 1941 that two Americans, Schaefer and Farrar, produced a pollen harvester consisting of a double grid of fine-mesh hardware cloth through which bees had to pass on returning to the hive. In so doing the pollen pellets were automatically removed from their legs and passed through a screen to the traps below. These traps were only used on a small scale, mainly to obtain supplies for fruit and crop pollination to aid the farmer or fruit grower. It will be appreciated that in those days no one had the slightest idea of how valuable pollen was to prove as a food and medicine. Honey, as always, was accepted, but even royal jelly was not recognized as being in any way superior to honey. The bee-keeper, scientists and health food experts had much more to learn about life in the beehive. Fortunately all this interesting information has now been obtained and man, as a result, is able to make fuller use of all the wonderful activities of the bee.

Life Saved by Pollen

Rose Wiseman, writing in the American publication *Herald of Health*, records a case of survival against all odds. In this instance an American soldier had been wounded and captured by the Chinese. Complete lack of medical aid resulted in his wounds becoming gangrenous until he could barely walk. His guards, seeing his helpless state, often left him unguarded. He managed to escape during the hours of darkness and limped or crawled till some friendly Chinese came to his aid. They gave him the best treatment possible and salves of pollen and honey were used as dressings on his wounds. He was also served various foods with pollen included. Some weeks later he was examined in an American hospital and the doctors were astounded at his wonderful recovery. It was considered that only a miracle had kept him alive. He was lucky, for that near-miracle in the form of pollen and honey saved his life.

Pollen in the Jungle

An even more remarkable illustration of the wonders of pollen as a survival food and life-saving medicine occurred during the Second World War. It was in 1941 that a Colonel Tretheway, a United States pilot, was captured by the Japanese after being shot down whilst on a raid. During the privations of the year which followed as a prisoner of war his weight was reduced from one hundred and seventy-five pounds to eighty-five pounds. At the camp there were almost daily executions of inmates no longer fit for work. The night before these terrible acts, selected prisoners were warned of their pending doom on the following day. The feet of all men due to be executed were seared with red-hot irons the night before to prevent their attempts to escape. The Colonel was subjected to this treatment but fortunately, with a comrade, managed to make good his escape. After the inhuman burns on his feet he was only able to make slow and agonizing progress through the jungle. He lost contact with his comrade, who was recaptured and shot. After months on a semi-starvation diet the weak man could now only exist on wild berries and shoots from the jungle. Additional hardship was caused when gangrene set into the wounds on his feet. Becoming completely exhausted and unable to crawl any further, he lay down to spend what he thought would be his last moments on earth and soon lost consciousness.

Pollen Eaters from the Jungle

On regaining consciousness he was amazed to find himself surrounded by members of a Chinese jungle tribe. They appeared friendly and as surprised to find him as he was to see them. He was carried on an improvised stretcher to their village, when he was subjected to awful agony as his feet were cleaned with brushes of stiff grass to remove the dead flesh. After being put in hot water, his body was carefully dried and the only treatment which followed consisted of the application of a thick coat of pollen to his feet followed by a layer of warm honey. These dressings were held in position by reed bandages which were changed every third day. This simple treatment proved successful and after four days he was relieved of the terrible pains in his feet. Two days later he was able to see his feet for the first time and to his surprise they were both intact, as were his toes. By now his swollen feet had been reduced to their normal size but were the colour of chocolate. For the first three days his only food consisted of a mealy substance which he later discovered was a mixture of pollen, with fruits, herb tea and broth. After this time he was also served honey, pollen cakes, fish, vegetables and meat.

This tribe, very backward by European or practically any standards, were able to obtain their supplies of pollen from all manner of waterways. An unlimited supply was available and could easily be fished out in nets especially made for this purpose. After the pollen was dried it was cleaned by the women, who used wooden tweezers to remove foreign bodies. The particles, some as large as peas, were then crushed up and stored in clay jars and sealed with mud. Such supplies could be used throughout the year when required, as a food, powdered medicine or antiseptic. The most health giving of foods and medicine was here produced in abundance by nature for the benefit of man. However, jungle-gathered pollens obtained by such crude methods cannot be compared to our products today. Modern Swedish pollen, gathered by the bees, is guaranteed to be absolutely pure and wholesome.

Pollen Cakes

Cakes were made by this Chinese tribe by spreading layers of sugared honey on a flat surface, with alternate fillings of pollen up to six thicknesses. This was then kneaded like dough. The mass, which then resembled a large pancake, was allowed to dry and then cut up into strips and left for about five days to completely dry out. These nourishing cakes were used as a survival food for hunters or in times of famine, drought, or during monsoon periods when other foods were hard to obtain.

The men of this tribe were tall, slender, and of healthy appearance: the women and children were also fine specimens. All enjoyed good health and young and old alike had sound teeth and were free from the aches and pains so prevalent in the 1980s.

Pollen – Perfect Food and Medicine

Pollen and honey as surgical dressing saved the life of Colonel Tretheway. As a food, this wonderful substance quickly restored him to a fine state of well-being not ever experienced before his captivity. In spite of his starved weak condition, gangrene-infested feet and the deprivations consequent on months in capitivity, within three weeks his feet were back to normal and he was able to walk once again. Even in well-equipped hospitals today gangrene usually means one has to suffer amputations – yet this man survived.

If a combination of pollen and honey can work miracles for a person at death's door, it will also do much for others who are happy, well-fed and in a reasonable physical state.

CHAPTER NINE

MEDICAL USES OF POLLEN AND ROYAL JELLY

As is always accepted in medical practice, new drugs and other medicines are first tried on animals. In the case of pollen it was shown to aid all kinds of creatures by improving fertility and promoting growth. During the course of various experiments to delay growth of tumours, animals have been given bee pollen in their feeds ranging from 1 part pollen to 10,000 parts food to 1 part pollen in 4,000 parts of food and this has proved successful in delaying the growth of various tumours.

From Animals to Humans

In many countries medical teams have made use of the advantages of both royal jelly and pollen in the treatment of varied complaints. A brief summary of some of the illnesses or defects, many of them quite serious, for which pollen will provide relief and in many cases effect a cure, include:

Premature ageing; Cerebral haemorrhage; Bodily weakness; Rickets; Anaemia; Loss of weight; Enteritis; Colitis; Toxic elimination; Constipation and many other conditions, some quite serious complaints, others as minor as the common cold.

These are only some of the illnesses, but still amount to quite a casebook of ailments which the average person might rightly have grave doubts about pollen being able to cure. However, if we refer to the vitamin, mineral, amino acids and other contents of pollen as discussed in Chapter One, we must accept it as being an amazing source of goodness.

Rutin

It has been mentioned that pollens, especially those from buckwheat, contain rutin. Now rutin generally appears only in wheat germ, so naturally pollen from wheat has a rich supply. Rutin is a glucoside and the finest food known for our arteries since it strengthens the entire system to resist such complaints as varicose veins. It will build a healthy heart, slow down its rhythm, enable blood to coagulate quicker and prevent bleeding. This little insight into only one component of pollen gives some idea how efficacious these tiny grains are for health and preventive medical treatment. It is no good waiting until signs of heart conditions or bulging varicose veins appear: the rutin in pollen, taken as a food supplement, will prevent such complaints.

Medical Opinions

Opinions of pollen as a medicine have been given by various doctors and some details are shown of the findings of these experts. All persons mentioned in case histories were organically fit, sound in mind and limb, but often very sick, in some cases both physically and mentally. So we shall see there are many and varied reasons for taking pollen and for each and every one these tiny grains have proved effective. What is most heartening is that it often effects cures where other treatments have failed. For the average person, not really ill but simply not enjoying robust health and energy, pollen will rapidly bring better health, with the happiness and optimism that goes with it.

Case Histories

Dr Rudolf Frey, Kornenburg Hospital, Austria, advocates pollen food for his patients. Four interesting cases dealt with by this doctor concerning persons between the ages of thirty-two and seventy are as follows:

A 35-year-old male in a very pessimistic and nervous condition due to overwork. Patient was organically fit but suffered insomnia and if not cured could become seriously ill. Within six days of taking pollen he became calm, relaxed and able to concentrate. Perfect sleep slowly returned. At the termination of the treatment his health was completely restored.

Male person 32-years-old suffering from continuous headaches and anxiety state. No physical abnormality and after taking pollen capsules for one week headaches ceased, nervous condition was no longer present and health returned to normal.

A 70-year-old man had always enjoyed good health except for the previous four months when he became very forgetful and unable to concentrate. He started to suffer from insomnia, with loss of energy. All treatment failed and his condition was simply attributed to old age. Many will wonder what can be done with a person aged seventy: the clock cannot be put back. However, many health foods, herbs, and other natural remedies will support the belief that you can do just this and pollen is one of them – perhaps the greatest – only time will tell. After the eighth day of treatment with pollen capsules, the patient made noticeable improvement and his powers of concentration increased considerably. By the end of the course he enjoyed a night's sleep once again and not only did pollen cure his complaints but the effects were lasting.

A female aged thirty-two had enjoyed perfect health except for the previous nine months when symptoms of vegetative dystonia with predominance of thyrogenous symptoms appeared. Hospital examination diagnosed an increase in the metabolic rate of 32 per cent. Patient being on night work for months had suffered from severe insomnia, exhaustion and the lack of concentration that accompanies it. Treatment with sedatives, sleeping tablets and tonics failed. After three weeks on pollen capsules her weight increased by four kilos and normal health was restored to such an extent that she could return to work.

Pollen and the Gynaecologist

Further testimony is given on the benefits of pollen by Dr Alois Schusta, a gynaecologist of Vienna, who outlines brief case histories of nine patients all of whom received two pollen capsules three times a day. Treatment continued for periods of eight days. Even after one or two eight-day courses the results were spectacular. The age range of these patients was from forty to fifty-nine, except for one person of twenty-nine. With these cases, as others, previous treatment had proved unsuccessful, except for temporary relief. Eight of the patients had climacteric or pre-climacteric complaints, withdrawal symptoms and disturbed sleep. These complaints, slow to respond to previous treatment, were quickly cured by pollen capsules. As with other illnesses the cure proved lasting even after the treatment ceased. Dr Schusta believes that the administration of a natural substance is far better than any synthetic medicine, especially as no side-effects or allergy can be caused by pollen.

Pollen Restores Lost Youth

Again, in Austria, from the Old People's Home in Lainz comes an interesting report by Dr Med. Franziska Stengel. This doctor used Polljuven (Austrian pollen preparation) in treating patients whose ages ranged from forty to eighty-five years. Three capsules were given with breakfast for seventeen days and the health of the persons rapidly improved. Even after eight days appetites had improved and weight started to increase; the gains being by the end of the course from one to four kilos per person. This was accompanied by a general all-round improvement in health, the patients being able to do more, sleep sounder and lead fuller lives. As in other cases, the benefits were retained even after the course of treatment was completed. Many very elderly persons received new vigour and energy, were able to walk greater distances and carry out tasks previously beyond their capabilities. As a result of these and other extensive tests, pollen is considered ideal for rejuvenating everyone, especially elderly people. Providing the body is organically healthy this treatment is splendid for convalescence, insomnia, lack of appetite, depression, the effects of stress, intestinal upsets, and so on.

Chronic Prostatitis

Since 1957 Cernitin pollen has been extensively used as a tonic in Sweden. During this long period it has proved effective for patients convalescing from operations or illness, and there have been no side-effects. In 1960 a Swedish specialist, Dr E. Ask-Upmark, Department of Medicine, University Hospital, Uppsala, issued reports that pollen would bring about a cure in a high percentage of cases of chronic prostatitis. This is a fairly common genital disease affecting male persons and which often proves difficult to treat. Two years later a Swedish specialist, Dr Leander, used pollen to treat one hundred cases of chronic prostatitis and nearly 80 per cent were cured. In 1967 Dr Ask-Upmark reported on twelve patients suffering from this illness who were again treated with Cernitin pollen: ten were completely cured.

The findings of these Swedish specialists have been confirmed by other experts. In East Germany Professor Helse endorsed the opinions of the Swedish doctors. Danish expert Professor Heise did likewise and also reported that pollen had quickly cured male patients suffering from various sexual problems. Pollen had proved effective in many cases when orthodox treatment was not only prolonged but often doomed to failure. Also from Germany, a Dr H. Klapsch, Director of a Department of Industrial Medicine, confirmed that AB Cernelle pollen tablets proved effective during an outbreak of influenza. By taking pollen 98 per cent of the workers suffering from this epidemic were able to continue working on heavy industrial production. This doctor was amazed at the success of pollen in keeping influenza at bay and it proved so much easier to administer than injections. In East Berlin three-year tests were conducted on male persons suffering from impotence, other sexual disorders, and chronic prostatitis. Many had a complete cure and were once again able to enjoy normal sex relations. Others recovered from prostatitis which had failed to respond to various kinds of treatment.

Cernitin and the Common Cold

A double blind test on colds and upper respiratory conditions treated with Cernitin extracts was carried out in northern Sweden on more than 700 troops during winter manoeuvres.

The findings showed that fewer soldiers in the group taking pollen reported sick because of upper respiratory tract infection than in the placebo group. Furthermore, soldiers who had the pollen extract were more alert and less tired than the others.

Pollen and Influenza

Pollen has also been tested against influenza and the results proved most satisfying. Following an outbreak of this illness, 459 employees of a heavy industrial plant in Germany were given pollen – the astounding result being that 98 per cent suffering from influenza were enabled to carry on working.

The Far East

In June 1967, report No. 8, volume 44 was issued by Dr Yutaka Saito, Department of Urology, Nagasaki School of Medicine, Japan. This deals with the treatment of chronic prostatitis with special reference to experience with Cernitin pollen. The findings made known in this report are most interesting, in that within a very short time over 80 per cent of the patients were cured. This illness, in spite of being a fairly common male complaint, has not always responded to treatment in the past and cures are often very prolonged. With chronic prostatitis diagnosis is somewhat complicated, so bearing this fact in mind, an even higher percentage of actual cures may have been effected from pollen treatment, as some patients, not classed as being cured, may not have been suffering from this complaint but from an illness showing similar symptoms.

Royal Jelly

Within the past twenty years royal jelly has been hailed in many countries as being superior to honey. Some doctors have endorsed this belief and there would seem to be more than a grain of truth in the assumption as indicated by the findings of scientists and other experts. Over a period of many years this jelly has been prescribed, with success, for a number of illnesses, these include:

Heart complaints. Persons with angina, infarction or other cardiovascular conditions obtained great help from royal jelly which was also shown to reduce cholesterol in the blood. It also brings the desired stabilizing effects of either increasing low blood pressure or reducing high blood pressure. For these varied conditions the amounts of royal jelly taken were about 100 mg. daily with honey for about thirty days.

CHAPTER TEN

THE FUTURE OF POLLEN

The actual number of doctors prescribing pollen in Sweden is not exactly known. However, some 4,000 medical practitioners have asked for samples, and many are obviously interested in the future of pollen for preventive treatment or as a medicine. For a long time the world's medical services have relied on drugs and ancient methods of healing naturally by herbs have gradually been abandoned. Various case histories have been mentioned in the previous chapter but other outstanding examples of 'Pollen Power' as a natural healer are briefly illustrated here:

Defies Medical Opinion

An unfortunate 34-year-old Swedish lady, Mrs Elmgaard, contracted a rare brain infection and was given only four months to live. She lost her sense of feeling and much weight, gradually becoming deaf, dumb and blind and was not given any hope of recovery. Fortunately, the timely intervention of AB Cernelle provided her with injections of pollen. She made a miraculous and almost complete recovery and once again enjoyed normal health. This case completely astounded medical men for the chances of this person surviving her terrible illness were remote to say the least. The facts of this case have been carefully checked by leading doctors and news of the miracle of pollen in curing Mrs Elmgaard appeared in many publications throughout the world, including *The Toronto Telegraph*, *Good Health*, Ireland's *Sunday Independent*, the Italian paper *Oggi Illustrato* and *Illustravana Politika* of Yugoslavia.

Another Life Saved

A Swedish pilot named Blahd was dangerously ill with severe burns after being involved in an air crash but was able to receive prompt treatment from Dr Per Sjostrom at the Engleholm Hospital. After his burns were cleaned pollen ointment was applied. He not only made a quick recovery but was soon able to fly again. Doctors were most surprised at the rapid success of this simple treatment in curing such terrible burns.

The two cases quoted here are outstanding, but nevertheless there would appear to be a definite future for pollen as indicated by the findings of a national newspaper. Evelyn Forbes, *Sunday Times* reporter, on a visit to Sweden observed pollen being prescribed by many doctors. After her fact-finding mission she wrote an article praising pollen as a natural remedy with no side-effects which gives nourishment to body cells and builds up resistance to disease. Evelyn Forbes was able to record successful results in pollen cures for many illnesses, including bronchitis, prostatitis, rheumatism and senility. She also noted that the Swedish Government, in addition to scientists and doctors, supported pollen as a healing agent after prolonged research and investigation. For example, Evelyn Forbes found that Oslo's Department of Social Hygiene had scientific evidence of the creative and rejuvenating power of pollen.

Natural Beauty Treatment

Natural foods such as olive oil, lemon juice, oatmeal and plants and herbs like the long forgotten stinging nettle provide excellent aids to fitness both internally and externally. Since man first gathered honey and bees-wax both have provided wonderful beauty preparations. The healing power of honey applies equally to the skin when used for burns, scalds and so on. Pollen, an even more remarkable substance, supplies a skin food and medicine which is even superior to honey. These little grains are being increasingly used in the cosmetic industry for they are free from the potentially dangerous chemicals which are usually employed. Swedish pollen supplies no less than seven beauty preparations in the form of lotions, creams, cleansing creams, powders, vanishing cream, skin and tissue cream, face lotions and hand treatment. There is a face lotion for the treatment of acne and skin infections; skin and tissue creams to counteract wrinkles. Swedish pollen can be obtained in all-purpose creams in combination with nutritive oils, fats and substances from the pollen of certain flowers which are ideal for offsetting premature ageing and promoting the growth of skin tissue. The simple difference in these beauty aids from nature is that pollen is alive, loaded with vitamins, amino acids and other goodness: as such it has a built-in power to bring skin back to life.

Conclusion

This booklet briefly introduces the health food and medicinal aspects of pollen, one of the most fascinating of subjects. These minute grains are available in all countries of the world. Countless millions of them are swept away in the breeze every day. Perhaps some of these masses of grains, which are far in excess of requirements for plant pollination, were provided for the use of man and animal alike.

Australia is one of the largest producers of honey in the world and yet at present there is only one bee-keeper known to the Australian Honey Board as being a collector of pollen; there being no known market available. Australian honey has become a household word, being as good as any and the most economical on the market. Pollen from Australia could eventually take on the same pattern, if a demand existed. It is there simply waiting collection from one of the finest honey-bee countries in the world. A market for pollen could start a second gold rush: this time for 'Golden Pollen Grains' from Australia, where the potential exists. Such an operation should equal, if not eventually surpass, all that has taken place in Sweden. In time this would bring the lasting benefits of pollen within reach of millions of people, for Australian bee-keepers, like their Swedish counterparts, would soon be able to gather pollen by the ton.

Critics may say that pollen is not meant for human consumption. Evidence accumulated during thirty years does not support this, for no side-effects have been reported from the use of pollen in its many forms. It has long been accepted that sprouting seeds such as wheat germ, nuts, peas and beans are wonderful for health and promote growth. As has been shown, these minute grains are present in varying amounts in all honeys and abundantly in royal jelly; all honey heaters having for years perhaps unwittingly consumed pollen. All the facts available show pollen, in its many forms, to be the greatest find from the world of the bee.

Pollen and the Law

In the Swedish courts the Consumer Ombudsman took out a summons against the Cernelle company for unsuitable advertising. The statement in dispute was: 'Increase efficiency with Cernelle pollen tablets.'

The decision of the courts made on the 16 July 1976 favoured the Cernelle company in these words: 'Summing up the considerations so set out, the Marketing Court finds therefore that the company has fulfilled reasonable demands for evidence of responsibility relative to the claim in its marketing.'